A GUIDE FOR ASPIRING DOCTORS

FROM *Cells*
TO SCALPELS

FIRST EDITION

WRITTEN BY AWARD-WINNING
AUTHOR YASMINE BEN SALMI

FIRST EDITION

FROM *Cells* TO SCALPELS

A GUIDE FOR ASPIRING DOCTORS

WRITTEN BY AWARD-WINNING
AUTHOR YASMINE BEN SALMI

FROM *Cells*
TO SCALPELS

A GUIDE FOR ASPIRING DOCTORS

**The Choice
is Yours**
PUBLISHING

Published by The Choice Is Yours Publishing

PAPERBACK ISBN: 978-1-915862-76-1
HARDBACK ISBN: 978-1-915862-75-4

FIRST EDITION

FROM *Cells* TO SCALPELS

A GUIDE FOR ASPIRING DOCTORS

WRITTEN BY AWARD-WINNING
AUTHOR YASMINE BEN SALMI

ACKNOWLEDGEMENTS

ACKNOWLEDGEMENTS

Writing a book is a journey, and I could not have reached this destination without the support and encouragement of the amazing people in my life. This book, "From Cells to Scalpels: A Guide for Aspiring Doctors," is the result of countless hours of research, reflection, and a burning desire to share my perspective and experiences with young individuals aspiring to join the medical field.

First and foremost, I want to express my deep gratitude to my family for their unwavering support throughout this project. Your belief in me and the passion I have for this subject matter has been a constant source of inspiration.

I would like to thank my mentors such as Dawn Kemp who is the Director for The Hunterian Museum here in London, who provided valuable insights as well as guidance while on my journey to becoming a plastic surgeon. Your wisdom and encouragement were instrumental in shaping the content and message of this work, and more to come.

I extend my appreciation to the countless healthcare professionals who generously shared their knowledge and experiences with the world to help better the medical industry as we know it today. Your dedication to the field of medicine and your willingness to impart your wisdom to the next generation of doctors is truly commendable.

I would like to acknowledge my family namely my mother Sabrina Ben Salmi, father Mohamed Ben Salmi, my siblings (Lashai Ben Salmi, Tray-Sean Ben Salmi, Paolo Ben Salmi and Amire Ben Salmi) and my grandmother Mary Paul for all their love, support and encouragement. My extraordinary mentors are Dawn Kemp - Director of Museums and Special Collections at the Royal College of Surgeons of England, Detina Zalli - Director of Pre-Medical Students at The University of Cambridge and Former Senior Lecturer at the University of Oxford and Harvard University, Lesley Warren - Senior Student Recruitment Officer (STEM) at the University of Brunel London and Rev Dr Trevor Adams - Director of The Black Hero.com whos also the access, Participation and Foundation Programme Leader at Christ The Redeemer College in Harrow. I would also like to acknowledge my associations for the unique opportunities that they pour into my life such as BABS aka the British Association of Black Surgeons and the ASGBI aka the Association of Surgeons of Great Britain and Ireland. I must acknowledge the young people like myself embarking on the journey toward a medical career. Your curiosity, ambition, and passion for making a difference in the world through medicine have motivated me to create this resource.

I also want to emphasize that this book is written from my perspective and personal experiences. It is not medical advice, as I am not a medical professional. Instead, it is intended to assist other young people in exploring their passion, questioning their motivations, and making informed decisions about their future in medicine. I hope that this book serves as a valuable tool in your journey of self-discovery.

Finally, I want to express my deepest gratitude to all the readers who have chosen to embark on this literary journey with me. I hope that "From Cells to Scalpels" provides you with insights, guidance, and inspiration as you navigate the path to becoming a doctor or finding your true calling within the medical field.

Sincerely,

Yasmine Ben Salmi

FIRST EDITION

FROM *Cells* TO SCALPELS

A GUIDE FOR ASPIRING DOCTORS

WRITTEN BY AWARD-WINNING
AUTHOR YASMINE BEN SALMI

DEDICATION

DEDICATION

This book is dedicated to all aspiring healers, whose journey into the medical field will learn from this book and treat it as a testament to the profound intersection of balance between science and compassion.

May your paths be illuminated by the pursuit of knowledge, the artistry of healing, and the unwavering commitment to making a difference in the lives of others.

This book is a tribute to the remarkable journey that lies ahead for each one of you who chooses to embark on a journey that has been proven to not be an easy one.

Hopefully, when I am all grown up, I will be able to look back on this book as a testament and realise just how much I have learned about life, people, empathy, joy, and lastly, what it really means to live a life of purpose.

FROM *Cells* TO SCALPELS

FIRST EDITION

WRITTEN BY AWARD-WINNING
AUTHOR YASMINE BEN SALMI

FIRST EDITION

FROM *Cells* TO SCALPELS

A GUIDE FOR ASPIRING DOCTORS

WRITTEN BY AWARD-WINNING
AUTHOR YASMINE BEN SALMI

INTRODUCTION

INTRODUCTION

Welcome to the World of Medicine

The allure of medicine: A lifelong fascination.

Imagine standing at the threshold of a realm where science, compassion, and resilience intersect. This is the world of medicine, a calling that has beckoned to countless individuals throughout history. From the days of ancient healers to the modern medical pioneers, the allure of medicine has drawn inquisitive minds into its embrace. For many, this fascination with the art and science of healing begins as a childhood dream, an innate curiosity that unfolds into a lifelong passion.

From the moment we take our first breath, our bodies become a canvas of life's intricate mysteries. Our hearts beat rhythmically, our lungs fill with air, and the symphony of biology plays out within us. Medicine provides the keys to decipher this symphony, to unlock the secrets hidden within the human body. It offers a journey of exploration and understanding, like no other.

As we embark on this journey from cells to scalpels, you'll soon come to realize that the allure of medicine lies not only in the extraordinary, in the remarkable cases that grace the pages of medical textbooks, but also in the ordinary moments of compassion and connection. It's in the reassuring smile you share with a worried patient, the comfort you bring to a family in their most trying times, and the lives you touch in immeasurable ways.

The journey ahead:

Challenges and rewards.
But make no mistake, this journey is not without its challenges.

The path to becoming a doctor is not a straight line; it's a winding road with its share of uphill climbs. The rigors of medical school, the pressure of long hours, the weight of life-and-death decisions, and the emotional toll of witnessing human suffering—all these will test your mettle. However, remember this: the challenges are the crucible in which you will be forged. They will shape you, strengthen you, and prepare you for the profound responsibility of healing. In the crucible, you'll discover your limits and then shatter them. You'll find resilience within yourself that you never knew existed.

You'll develop a profound understanding of the human condition, one that can only be gained through experience. The rewards, too, are boundless. There's the satisfaction of knowing you've made a difference, the privilege of earning your patients' trust, and the exhilaration of contributing to the ever-evolving tapestry of medical knowledge. It's in the moments when you witness a life being brought back from the brink, or when you provide solace to a patient in their final moments. These moments, these rewards, make every challenge worthwhile.

FIRST EDITION

FROM *Cells* TO SCALPELS

A GUIDE FOR ASPIRING DOCTORS

WRITTEN BY AWARD-WINNING
AUTHOR YASMINE BEN SALMI

CHAPTER 1

CHAPTER 1

The Human Body - A Masterpiece of Engineering

In the heart of every aspiring doctor, there is a relentless curiosity, an insatiable desire to understand the intricate mechanisms that power the human body. To embark on a journey from cells to scalpels is to enter a realm of wonder and discovery, where science and compassion converge. Welcome to the world of medicine, where the excitement of becoming a doctor intertwines with the profound mysteries of the human body.

As we begin our journey, let us first consider the human body, a masterpiece of engineering. It is a harmonious blend of biological complexity and remarkable precision, a symphony of cells, tissues, organs, and systems working together in unison. At the core of this symphony are the trillions of cells that form the building blocks of life.

Picture a world so small that it exists beyond the range of our naked eyes. In this miniature universe, cells are the architects, engineers, and labourers. They construct the foundation upon which the body is built, and from them, all life emanates. These microscopic entities are the true marvels of biology, each with its own unique role, yet all cooperating to create the vibrant tapestry of human existence.

From the bustling cities of neurons in our brains to the ceaseless traffic of blood cells through arteries and veins, every part of our body plays a vital role in maintaining our health. The human body is not only a physical wonder but also a biological puzzle waiting to be solved. This enigma is what beckons the curious and the compassionate to the world of medicine.

Medicine is not just a science; it is an art. It requires the blending of knowledge and empathy, the ability to peer into the intricate machinery of life and recognize when it falters. It is the understanding that each patient is unique, and their stories and struggles are as important as their symptoms. A doctor's journey begins with a fascination for the human body but is incomplete without a deep appreciation for the individuals who inhabit it.

The allure of medicine lies in its transformative power. It allows you to witness the birth of a new life, mend the broken, and ease the suffering. It's a discipline that touches the very essence of humanity, where the act of healing extends far beyond the physical realm. It's a privilege that comes with great responsibility, as doctors hold in their hands the well-being and sometimes even the lives of their patients.

As you venture further into the world of medicine, you will explore the complexities of human biology, the power of diagnosis, and the ever-evolving field of medical technology. You will learn to unravel the mysteries of diseases, understand the role of genetics, and grapple with ethical dilemmas that accompany the practice of medicine.

This journey is not without its challenges, but the rewards are immeasurable. While reading from cells to scalpels, you will discover a world where the line between science and wonder blurs, and where every patient's story adds a unique chapter to your own.

So, fasten your seatbelt, and prepare to embark on a journey through the fascinating world of medicine. In the chapters that follow, we will delve deeper into the realms of the human body, exploring its remarkable systems, the tools of the medical trade, and the dedication it takes to become a healer. Welcome to the world of medicine, where curiosity meets compassion, and the human body is the canvas on which we paint our dedication and care.

"Medicine is not only a science; it is also an art. It does not consist of compounding pills and plasters; it deals with the very processes of life, which must be understood before they may be guided." — Paracelsus

CHAPTER 2

CHAPTER 2

The Human Body - The marvels of human anatomy and physiology

In the vast cosmos of the medical world, there exists no grander spectacle than the human body itself. As we delve into the intricacies of anatomy and physiology, we uncover a masterpiece of engineering —a complex, awe-inspiring creation.

Our journey into medicine begins with an exploration of the marvels that exist within each of us. The human body, a work of art sculpted by the forces of evolution, is a symphony of cells, tissues, and organs, each playing its unique role in sustaining life.

From the microscopic intricacies of DNA to the grandeur of the musculoskeletal system, every facet of the human body serves a purpose. As a future doctor, you'll be tasked with unravelling the enigma of human biology. The more you learn, the more you'll appreciate the genius that is the human body. From the brain's remarkable neural network to the resilient and adaptable immune system, each system holds its own set of wonders.

Systems at work: How the body functions as a whole

The beauty of the human body is not only in its individual components but also in the way these components work in harmony. It's a symphony of life, and understanding the interplay between its various sections is crucial for anyone aspiring to be a doctor.

The cardiovascular system, with its intricate network of blood vessels and the rhythmic pumping of the heart, ensures the distribution of life-giving oxygen and nutrients.

The respiratory system, with its rhythmic exchange of air in and out of the lungs, provides the oxygen needed for cellular respiration.

The digestive system transforms food into energy, while the nervous system orchestrates the body's functions and coordinates responses to external stimuli.

This interconnectedness serves as a constant reminder that the body is more than just the sum of its parts. To heal, to understand, and to excel in medicine is to comprehend not only the individual systems but also their integration into a harmonious whole.

"The human body is the best picture of the human soul."
— Ludwig Wittgenstein

FIRST EDITION

FROM *Cells*
TO SCALPELS

A GUIDE FOR ASPIRING DOCTORS

WRITTEN BY AWARD-WINNING
AUTHOR YASMINE BEN SALMI

CHAPTER 3

CHAPTER 3

Discovering Your Passion for Medicine

The path to becoming a doctor is not a one-size-fits-all journey. It is a unique voyage, and in this chapter, we explore how to discover your personal passion for medicine. As you embark on this adventure, it's essential to remember that your enthusiasm, interests, and motivations are the compass guiding you through the world of medicine.

The Seed of Curiosity

For many aspiring doctors, their fascination with medicine often begins at an early age. Perhaps you've always been intrigued by the human body and its functions, or maybe you've been inspired by a family member who works in healthcare.

It might have started with a simple question: "How does the body work?" This initial spark of curiosity is the foundation upon which your passion for medicine can grow.

The journey into medicine often takes shape through experiences that stoke that curiosity. It may be volunteering at a local clinic, shadowing a physician, or engaging in science projects that delve into the intricacies of life. These hands-on experiences can help you determine whether medicine is the right path for you.

Exploring Different Specialties

Medicine is a vast field with a wide range of specialities, from paediatrics to surgery, psychiatry to radiology. Each speciality offers its unique challenges and rewards. To discover your passion for medicine, it's important to explore these diverse options.

Consider what aspects of medicine resonate with you. Are you drawn to the immediate and life-saving interventions of the emergency room, or are you more inclined toward the long-term relationships of primary care? The more you explore these specialities, the clearer your own interests will become.

Mentorship and Guidance

One invaluable resource in your journey to discover your passion for medicine is mentorship. Seek out experienced healthcare professionals who can guide you, share their insights, and provide you with a broader perspective on the field. A mentor can offer invaluable advice, helping you understand the challenges and rewards of a medical career.

Personal Values and Compassion

Medicine is not just about scientific knowledge; it's also about compassion and a commitment to the well-being of others. Reflect on your own values and what drives you to pursue a career in medicine. Are you passionate about helping people, making a difference in the world, or advancing medical research? Understanding your underlying motivations will strengthen your dedication to the field.

Overcoming Challenges

Becoming a doctor is a challenging journey, and there will be obstacles along the way. These challenges, whether academic, personal, or professional, can test your passion for medicine. It's during these moments of struggle that you'll need to tap into your resilience and unwavering commitment to your chosen path.

Continuous Learning

Medicine is a dynamic field that is constantly evolving. To maintain your passion, you must be open to lifelong learning. Stay curious and up-to-date with the latest advancements in medical science, as this ongoing education will not only deepen your understanding but also fuel your passion.

In the world of medicine, discovering your passion is an ongoing process. Your journey may take unexpected turns, but it's the continuous exploration, growth, and commitment to helping others that will define your path. As you navigate your way through the world of medicine, remember that your unique passion and perspective are valuable assets in this noble profession. Whether your journey started as a childhood curiosity or a newfound interest, your passion for medicine is the key to making a meaningful impact on the lives of others.

"The two most important days in your life are the day you are born and the day you find out why."
— Mark Twain

FIRST EDITION

FROM *Cells*
TO SCALPELS

A GUIDE FOR ASPIRING DOCTORS

WRITTEN BY AWARD-WINNING
AUTHOR YASMINE BEN SALMI

CHAPTER 4

CHAPTER 4

The Journey to Medical School

The path to medical school is a challenging and rewarding odyssey that requires dedication, perseverance, and unwavering commitment. In this chapter, we will explore the steps and considerations that make up the journey to medical school, from the early preparations to the moment you step into that hallowed institution.

Preparation: Building a Strong Foundation

Your journey to medical school begins well before you submit your application. It starts with building a strong academic foundation. High school and undergraduate studies provide the basis for your future medical education. While there isn't a specific major required for medical school, most students pursue degrees in biology, chemistry, or related fields. Maintaining a high GPA is crucial, as it demonstrates your academic capabilities.

Extracurricular Activities and Volunteer Work

In addition to academics, medical schools look for well-rounded individuals who have demonstrated a commitment to their communities and a genuine passion for helping others. Participating in extracurricular activities, volunteering in healthcare settings, and engaging in research projects can set you apart from other applicants. These experiences allow you to develop crucial skills, such as empathy, teamwork, and leadership, which are essential for a career in medicine.

The MCAT: Gateway to Medical School

The Medical College Admission Test (MCAT) is a significant milestone on your journey to medical school. This standardized exam assesses your knowledge of the natural sciences, critical thinking, and problem-solving abilities. Proper preparation is key, and many aspiring medical students invest months in rigorous study programs to achieve a competitive score.

Choosing the Right Medical Schools

Once you've met the academic and extracurricular requirements, it's time to choose the medical schools to which you'll apply. Consider factors such as location, class size, specialities offered, and the school's mission and values. Applying to a mix of reach, match, and safety schools increases your chances of acceptance.

The Application Process

The primary application for medical school in the United States is submitted through the American Medical College Application Service (AMCAS) or the American Association of Colleges of Osteopathic Medicine Application Service (AACOMAS) for osteopathic medical schools. The application typically includes transcripts, letters of recommendation, a personal statement, and information about your extracurricular activities and work experience.

Interviews and Secondary Applications

If your primary application is successful, you may be invited for interviews at medical schools. Interviews are an opportunity for the admissions committee to get to know you on a personal level. It's essential to prepare for interviews, not only by practicing common interview questions but also by reflecting on your motivations for pursuing a medical career.

Acceptance and Decision-Making

Getting accepted into medical school is a momentous achievement, but it's not the end of your journey. Once you receive acceptance, you'll need to consider factors such as location, cost, and the curriculum when making your final decision. Your chosen medical school will be your home for the next several years, so it's crucial that it aligns with your goals and values.

Financial Planning

Medical education is expensive, and it's essential to plan for the financial aspects of your journey. Scholarships, grants, and student loans are some of the ways to fund your medical education. Research and apply for financial aid opportunities early in the process to help alleviate the financial burden.

The White Coat Ceremony

The culmination of your journey to medical school is often marked by the White Coat Ceremony, a symbolic event where you, as a student doctor, don the white coat that represents your commitment to the profession. It is a moment of pride, achievement, and the beginning of your formal medical training.

The journey to medical school is a challenging but ultimately rewarding path.

It's a testament to your passion, commitment, and your desire to make a difference in the world through the practice of medicine. As you move forward, always remember why you embarked on this journey in the first place, and let that motivation guide you through the challenges and triumphs that lie ahead.

"It is not the strongest of the species that survive, nor the most intelligent, but the one most responsive to change."
— Charles Darwin

CHAPTER 5

CHAPTER 5

Anatomy 101 - The Building Blocks of Life

Anatomy is the cornerstone of medicine. It is the study of the structure of the human body, a discipline that forms the foundation of our understanding of health, disease, and the practice of medicine. In this chapter, we'll delve into Anatomy 101, where we explore the intricate organization of the human body, from the smallest building blocks to the complex systems that sustain life.

The Cellular Blueprint

At the core of human anatomy are cells, the fundamental units of life. Cells come in various types, each uniquely adapted to its function. From the specialized neurons in the brain to the contractile muscle cells in the heart, every cell type plays a vital role in maintaining the body's functions.

The cell's nucleus contains the genetic material, DNA, which carries the instructions for the cell's activities. This genetic code is the blueprint for building proteins, enzymes, and other essential molecules that regulate our bodily functions.

Tissues: Cells Working Together

Cells are rarely alone; they collaborate to form tissues, the next level of anatomical organization. There are four primary types of tissues: epithelial, connective, muscle, and nervous. Epithelial tissues cover surfaces, protect underlying structures, and serve as barriers. Connective tissues support, protect and bind various parts of the body. Muscle tissues enable movement, while nervous tissues transmit information through electrical impulses.

Organs: Complex Structures with Specific Functions

Tissues combine to form organs, which are structures with distinct functions. Examples include the heart, lungs, liver, and brain. Each organ performs essential roles in maintaining the body's overall health and vitality. Organs often contain multiple tissue types that work together, such as the heart, which consists of muscle, connective, and nervous tissues.

Systems: Coordinated Networks of Organs

Organs do not function in isolation but work together within systems to regulate specific bodily processes. The cardiovascular system, for instance, consists of the heart, blood vessels, and blood, working collaboratively to transport oxygen and nutrients to cells throughout the body.

Some of the major organ systems in the human body include:

Circulatory System:
Responsible for blood circulation, supplying oxygen and nutrients.

Respiratory System:
Involved in breathing and gas exchange.

Nervous System:
Controls communication through electrical impulses.

Digestive System:
Processes and absorbs nutrients from food.

Endocrine System:
Manages hormones and chemical signaling.

Muscular System:
Enables movement and posture.

Skeletal System:
Provides structural support and protects organs.

Immune System:
Defends the body against pathogens.

Reproductive System:
Facilitates reproduction and the continuation of the species.

Dissection and Exploration

The study of anatomy often involves dissection, the careful examination and disassembly of cadavers to explore the body's structures. This hands-on experience allows students and medical professionals to gain a deep understanding of the body's complexities. Modern technology, such as medical imaging and virtual dissection, has also expanded our ability to study anatomy.

Clinical Significance of Anatomy

Understanding anatomy is critical for medical practitioners. It provides the knowledge needed to diagnose and treat patients. A comprehensive grasp of anatomy enables doctors to locate and understand the significance of injuries, diseases, and abnormalities in the body.

A Lifelong Journey

The study of anatomy is not confined to medical school but is a lifelong journey for healthcare professionals. It evolves as new discoveries are made, and it deepens with clinical experience.

As you progress in your medical education, you will continue to explore the intricacies of the human body, appreciating how its structures and systems interconnect to sustain life. This foundation in anatomy is not just knowledge; it's a profound appreciation of the incredible complexity of the human body and the beginning of your journey as a physician.

"The body is a sacred garment. It is what you enter life in and what you depart life with, and it should be treated with honor."
— Martha Graham

CHAPTER 6

CHAPTER 6

The Art of Healing - Medicine's Human Side

Medicine is often described as a science, but it is equally an art. This chapter delves into the human side of medicine, emphasizing the vital role that empathy, compassion, and effective communication play in the practice of healthcare.

The Human Connection

At the heart of medicine is the relationship between healthcare providers and their patients. The bond formed in this partnership is a source of strength, comfort, and trust. It's where the art of healing truly begins.

Empathy and Compassion

Empathy is the ability to understand and share the feelings of another. In the medical context, empathy is the key to connecting with patients on a deeply personal level. It enables healthcare providers to comprehend their patients' physical and emotional pain, fears, and anxieties. By demonstrating empathy, healthcare professionals not only diagnose and treat illnesses but also provide solace and support during difficult times.

Compassion, on the other hand, is the genuine desire to alleviate suffering and improve the well-being of patients. It's the force that drives doctors and nurses to go above and beyond in their care. Compassion involves acts of kindness, understanding, and advocacy on behalf of patients, ensuring they receive the best possible care.

Effective Communication

Clear and effective communication is the cornerstone of patient care. Healthcare providers must be adept at conveying complex medical information in a way that patients can understand. Listening actively and attentively to patients' concerns is just as important as relaying information. It is through communication that trust is established, questions are answered, and informed decisions are made.

The Power of a Healing Touch

The human touch is a powerful tool in medicine. Physical contact can comfort, console, and reassure patients. A reassuring hand on the shoulder or a gentle touch during an examination can make a world of difference. Touch is a language of empathy, a way to convey caring and support without words.

Ethical Dilemmas and Tough Decisions

The practice of medicine is not without its ethical dilemmas. Healthcare providers often find themselves in situations where they must balance competing interests, such as patient autonomy, beneficence, and justice. The art of healing involves navigating these complexities with integrity and sensitivity.

The Healing Environment

Hospitals and medical facilities play a significant role in the art of healing. Creating a healing environment goes beyond sterile walls and state-of-the-art equipment. It encompasses a compassionate and patient-centered atmosphere where the physical space itself contributes to the well-being of patients.

Cultural Sensitivity and Diversity

Cultural competence is a vital aspect of the art of healing. Patients come from diverse backgrounds with unique values and belief systems. Understanding and respecting these differences are essential for providing effective care. Culturally sensitive healthcare acknowledges the importance of diversity and fosters an inclusive environment.

The Continuous Pursuit of Excellence

The art of healing is a lifelong pursuit. Healthcare professionals continually seek opportunities to improve their skills, enhance their empathy, and deepen their understanding of patient needs. It is a commitment to not only practising medicine but practising it with a profound commitment to the well-being of patients.

Take away from this chapter

As you progress in your medical journey, remember that the art of healing is not a separate entity from the scientific side of medicine; it is an integral part of the whole. The most effective healthcare providers blend medical knowledge with compassion and empathy, ultimately delivering care that addresses not only the physical aspects of illness but the emotional and spiritual needs of the patient. This balance between science and humanity is what defines the noble profession of medicine.

"Kind words do not cost much. Yet
they accomplish much."
— Blaise Pascal

FIRST EDITION

FROM *Cells* TO SCALPELS

A GUIDE FOR ASPIRING DOCTORS

WRITTEN BY AWARD-WINNING
AUTHOR YASMINE BEN SALMI

CHAPTER 7

CHAPTER 7

Beyond the Textbooks - Cutting Edge Medical Research

Medicine is a field that is in a constant state of evolution, with researchers and scientists pushing the boundaries of knowledge to improve patient care, develop new treatments, and enhance our understanding of the human body. In this chapter, we explore the world of cutting-edge medical research and its impact on the practice of medicine.

The Thrill of Discovery

Medical research is a dynamic and ever-evolving field. It's a quest for knowledge, innovation, and discovery. Researchers are driven by the desire to unravel the mysteries of the human body, tackle pressing health issues, and find solutions that will benefit patients worldwide.

Translational Research

Translational research bridges the gap between basic science and clinical applications. It takes the discoveries made in laboratories and translates them into practical solutions for patient care. This process involves a collaborative effort between researchers, physicians, and healthcare institutions to ensure that scientific breakthroughs are applied to real-world medical challenges.

Key Areas of Research

Cutting-edge medical research encompasses a wide range of areas, including:

Genomics and Precision Medicine:

The study of an individual's genetic makeup to tailor treatments and interventions to their specific needs. Genomics has led to breakthroughs in cancer therapy, rare disease diagnosis, and pharmacogenomics.

Regenerative Medicine:

Exploring the use of stem cells, tissue engineering, and other approaches to repair or replace damaged or diseased tissues and organs. This research holds promise for conditions like heart disease, spinal cord injuries, and diabetes.

Neuroscience:

Understanding the brain and the nervous system to advance treatments for neurological and psychiatric disorders, including Alzheimer's disease, Parkinson's disease, and depression.

Immunotherapy:

Harnessing the body's immune system to target and destroy cancer cells. Immunotherapy has revolutionized cancer treatment and holds great potential for the future.

Global Health and Infectious Diseases:

Addressing health disparities and tackling infectious diseases, such as HIV/AIDS, malaria, and emerging pathogens like the Zika virus.

Artificial Intelligence and Big Data:

Utilizing advanced technology to analyze vast amounts of medical data for improved disease diagnosis, prediction, and treatment.

Clinical Trials and Evidence-Based Medicine

Clinical trials are a crucial component of medical research. These trials test the safety and efficacy of new drugs, treatments, and interventions. Evidence-based medicine relies on the results of clinical trials and other rigorous research to guide medical practice. It ensures that healthcare decisions are based on the most reliable and up-to-date information available.

The Role of Medical Professionals in Research

Medical professionals are not just practitioners; they are also contributors to medical research. Many clinicians are actively involved in research projects, often at academic medical centres. These professionals bridge the gap between the laboratory and the clinic, applying their findings to patient care.

Challenges and Ethical Considerations

Medical research is not without its challenges. Ethical considerations, patient safety, and the responsible use of data are of paramount importance. Balancing innovation with the potential risks of untested treatments is an ongoing challenge.

Patient Participation

Patients play a crucial role in medical research by participating in clinical trials and studies. Their involvement is vital in advancing medical knowledge and developing new treatments. Patient advocacy and support groups also contribute to research efforts.

The Future of Medical Research

The future of medical research holds great promise. Advances in technology, interdisciplinary collaboration, and the increasing globalization of research efforts are shaping the landscape of medicine. Medical researchers continue to explore new frontiers, seeking to improve patient care, enhance our understanding of disease, and find cures for conditions that have long eluded us.

As you embark on your own journey in the field of medicine, remember that you have the opportunity to not only apply the latest research findings in your clinical practice but also to contribute to the advancement of medical knowledge. Your involvement in cutting-edge medical research can make a significant impact on the lives of patients and the future of healthcare.

"Research is creating new knowledge."
— Neil Armstrong

CHAPTER 8

CHAPTER 8

Specializations and Superheroes

In the world of medicine, healthcare professionals don various capes, not made of cloth but of knowledge, expertise, and dedication. They are the superheroes of modern times, each with their unique specializations, coming together to provide comprehensive care to patients. In this chapter, we explore some of the diverse medical specializations and the superheroes who pursue them.

1. Cardiologists - The Heart's Guardians

Cardiologists are the guardians of the heart, specializing in the diagnosis and treatment of heart and vascular conditions. They monitor heart health, perform life-saving interventions, and help patients manage heart disease risk factors.

2. Neurologists - The Brain Whisperers

Neurologists are the brain whisperers, delving into the complexities of the nervous system. They diagnose and treat neurological conditions such as strokes, epilepsy, Alzheimer's disease, and multiple sclerosis.

3. Surgeons - The Precision Craftsmen

Surgeons are the precision craftsmen of medicine. Whether in the operating room or in a specialized field like orthopaedics, plastic surgery, or neurosurgery, they use their hands to heal and restore.

4. Pediatricians - The Children's Champions

Paediatricians are the champions of children's health. They provide comprehensive care to infants, children, and adolescents, ensuring their growth, development, and well-being.

5. Obstetricians and Gynecologists - The Bringers of Life

Obstetricians and gynaecologists bring life into the world. They care for women during pregnancy, childbirth, and beyond, addressing reproductive health and wellness.

6. Psychiatrists - The Mind Healers

Psychiatrists are the mind healers, helping individuals cope with mental health challenges. They provide therapy, medications, and support for conditions such as depression, anxiety, and schizophrenia.

7. Oncologists - The Cancer Warriors

Oncologists are the warriors against cancer. They diagnose and treat various forms of cancer, often combining surgery, chemotherapy, and radiation therapy to combat the disease.

8. Emergency Medicine Physicians - The First Responders

Emergency medicine physicians are the first responders to life-threatening situations. They remain calm under pressure, making rapid decisions to save lives in the chaotic environment of the emergency department.

9. Anesthesiologists - The Comfort Providers

Anesthesiologists ensure comfort and safety during surgeries and medical procedures. They administer anaesthesia, monitor vital signs, and manage pain.

10. Radiologists - The Diagnostic Wizards

Radiologists are the diagnostic wizards of medicine. They interpret medical images, such as X-rays, CT scans, and MRIs, to aid in the diagnosis and treatment of a wide range of conditions.

11. Pathologists - The Disease Detectives

Pathologists are disease detectives, uncovering the secrets of diseases through the study of tissues, cells, and laboratory tests. Their findings are crucial for accurate diagnoses.

12. Geriatricians - The Elders' Advocates

Geriatricians are advocates for the elderly. They specialize in the unique healthcare needs of older adults, addressing aging-related conditions and providing quality-of-life care.

13. Family Medicine Physicians - The Health Coordinators

Family medicine physicians are health coordinators for families and individuals of all ages. They provide comprehensive care, addressing a wide range of medical needs.

14. Infectious Disease Specialists - The Microbe Warriors

Infectious disease specialists are the warriors against infectious microbes. They diagnose and treat infectious diseases, and they often play a vital role in public health.

15. Allergists and Immunologists - The Allergy Detectives

Allergists and immunologists are the detectives who uncover the causes of allergies and immune system disorders, helping patients manage their conditions.

16. Nephrologists - The Kidney Experts

Nephrologists are the kidney experts, specializing in the diagnosis and treatment of kidney diseases and related conditions.

17. Pulmonologists - The Breathing Specialists

Pulmonologists are specialists in breathing, and treating lung conditions such as asthma, chronic obstructive pulmonary disease (COPD), and more.

18. Endocrinologists - The Hormone Balancers

Endocrinologists are experts in hormones, helping patients manage conditions related to the endocrine system, such as diabetes and thyroid disorders.

19. Rheumatologists - The Joint Care Experts

Rheumatologists specialise in caring for patients with rheumatic conditions, including arthritis and autoimmune diseases that affect the joints.

20. Gastroenterologists - The Digestive Health Advocates

Gastroenterologists are advocates for digestive health. They diagnose and treat disorders of the digestive system, including the stomach, intestines, liver, and pancreas.

Each medical speciality represents a unique set of skills, knowledge, and expertise. While these healthcare professionals may not wear capes, they are heroes in their own right, dedicated to the well-being of patients and making a positive impact on the world of medicine. As you explore the various medical specializations, you will find the one that resonates with your passion and allows you to become a superhero in your own unique way.

"Specialization is for insects."
— Robert A. Heinlein

CHAPTER 9

CHAPTER 9

Trials and Triumphs - The Life of a Doctor

The life of a doctor is a remarkable journey filled with trials, challenges, and triumphant moments. In this chapter, we explore the realities and experiences that define the path of a physician, from the demanding years of medical training to the deeply rewarding moments of patient care.

The Journey of Medical Education

A doctor's path begins with rigorous education and training. Medical school is a demanding period of intense study, long hours, and clinical rotations. The curriculum covers a wide range of medical subjects, from anatomy and pharmacology to ethics and patient communication. The journey is challenging, requiring dedication, resilience, and a deep passion for medicine.

Residency - The Crucible of Clinical Training

After medical school, new doctors enter residency programs, which provide hands-on clinical experience under the guidance of experienced mentors. Residencies vary by speciality and can last from three to seven years. This is a time of intensive learning, during which residents develop their clinical skills, make life-altering decisions, and often endure sleepless nights.

Board Examinations and Licensure

In addition to their training, doctors must pass board examinations to become licensed in their chosen speciality. These exams are rigorous and require thorough preparation. Successfully becoming board-certified is a significant milestone in a doctor's career.

Balancing Work and Life

The demanding nature of medical training and practice often presents a challenge in achieving a work-life balance. Doctors must learn to manage their time effectively, maintain their physical and mental health, and nurture their personal lives and relationships.

Patient Care - The Core of Medicine

The true essence of a doctor's life lies in patient care. Doctors form deep bonds with their patients, helping them navigate illness, injury, and the complexities of the healthcare system. The ability to make a difference in the lives of those they care for is one of the most rewarding aspects of the profession.

Trials and Challenges

The life of a doctor is not without trials and challenges. They may face difficult ethical decisions, medical errors, and the emotional toll of witnessing suffering. These challenges can be emotionally draining but also serve as opportunities for growth and reflection.

Medical Innovations and Advancements

Doctors are at the forefront of medical innovations and advancements. They contribute to the development of new treatments, surgical techniques, and healthcare technologies. Staying informed about the latest research and applying it in clinical practice is a dynamic aspect of the profession.

A Lifelong Journey of Learning

Medicine is a lifelong journey of learning. Doctors must continually update their knowledge, adapt to new healthcare practices, and embrace the ever-evolving field of medical science.

Patient Relationships and Gratitude

The relationships formed between doctors and their patients are a source of immense gratification. Doctors often witness the transformation of patients from illness to recovery, and these moments of triumph are the rewards for their dedication.

The Oath and the Duty

The Hippocratic Oath, taken by most medical professionals, encapsulates the sense of duty and ethics that guide their actions. It serves as a reminder of their responsibility to prioritize patient well-being and uphold the highest standards of medical practice.

Legacy and Impact

A doctor's legacy extends beyond their individual practice. They contribute to the betterment of society by improving the health and well-being of countless individuals. The impact of their work is immeasurable.

The life of a doctor is one of commitment, compassion, and resilience. It is a journey marked by both trials and triumphs, where the pursuit of medical excellence is balanced with the duty to care for patients. The life of a doctor is a calling that transforms passion into a profession and service into a lifelong commitment. As you progress on your own path in the world of medicine, remember that the challenges you face and the triumphs you achieve are all part of a noble and impactful journey.

"Success is not final, failure is not fatal: It is
the courage to continue that counts."
— Winston Churchill

FIRST EDITION

FROM *Cells* TO SCALPELS

A GUIDE FOR ASPIRING DOCTORS

WRITTEN BY AWARD-WINNING AUTHOR YASMINE BEN SALMI

CHAPTER 10

CHAPTER 10

Preparing for Med School - Tips and Resources

The path to medical school is a challenging yet rewarding journey. In this chapter, we provide tips and resources to help you prepare effectively for medical school and increase your chances of success.

1. Early Preparation:

- **Build a Strong Foundation:** Start by excelling in your high school and undergraduate courses, especially in science and math. A solid academic foundation is crucial for medical school admissions.

- **Extracurricular Activities:** Engage in extracurricular activities, such as volunteering, research, and leadership roles, to demonstrate your commitment to healthcare and your ability to balance multiple responsibilities.

2. Preparing for the MCAT:

- **Start Early:** The MCAT (Medical College Admission Test) is a critical part of your application. Begin preparing well in advance to maximize your score.

- **Review Resources:** Utilize study materials, practice exams, and review courses to familiarize yourself with the test format and content.

3. Clinical Experience:

- **Volunteer or Work in Healthcare:** Gain hands-on clinical experience by volunteering or working in hospitals, clinics, or healthcare facilities. This demonstrates your commitment to patient care.

4. Research:
- **Engage in Research:** If you have an interest in medical research, seek opportunities to participate in scientific projects. Research experience can enhance your application.

5. Letters of Recommendation:
- **Cultivate Relationships:** Build strong relationships with professors, supervisors, and mentors who can write compelling letters of recommendation on your behalf.

6. Extracurriculars:
- **Quality Over Quantity:** Rather than being involved in numerous activities, focus on a few that align with your interests and allow you to make a meaningful impact.

7. The Application Process:
- **Plan Early:** Begin the application process early and thoroughly review the requirements of each medical school to which you intend to apply.

- Write a Strong Personal Statement: Your personal statement should be a compelling narrative that highlights your passion for medicine, experiences, and your commitment to the field.

8. Financial Considerations:
- **Financial Planning:** Investigate financial aid options, scholarships, and loan programs to ensure you can manage the costs of medical school.

9. Interview Skills:

- **Practice Interviewing:** Develop your interviewing skills through mock interviews and self-assessment. Be prepared to discuss your experiences, motivations, and goals confidently.

10. Stay Informed:

- **Keep up with Medical News:** Stay informed about current events and medical breakthroughs. This demonstrates your genuine interest in the field.

11. Maintain a Support System:

- **Seek Support:** Lean on mentors, family, and friends for encouragement and guidance throughout the application process.

- **12. Self-Care:**
- **Prioritize Your Health:** Maintain a healthy lifestyle, exercise regularly, and manage stress to ensure you are physically and mentally prepared for the challenges of medical school.

13. Gap Year Considerations:

- **Consider a Gap Year:** Taking a gap year can provide you with valuable life experiences, additional time to prepare, and a chance to enhance your application.

Resources:

- Pre-Medical Advisors: Consult your college's pre-medical advisors for guidance on course selection, extracurricular activities, and the application process.

- MCAT Prep Courses: Enroll in MCAT prep courses or use official AAMC resources to prepare for the MCAT.

- Medical School Admission Services: Explore resources provided by the Association of American Medical Colleges (AAMC) and the American Association of Colleges of Osteopathic Medicine (AACOM) to learn about the application process.

- Online Communities: Join online forums and communities where pre-medical and medical students share advice, experiences, and support.

- Medical School Admissions Books: Numerous books provide comprehensive information on the medical school application process, interviews, and personal statements.

- Medical School Websites: Visit the official websites of the medical schools to which you plan to apply. They often provide detailed information about their admission requirements and processes.

- Medical School Interviews: Consider resources and books that focus on medical school interview preparation and etiquette.

The journey to medical school is a significant undertaking, but with careful planning, dedication, and the use of available resources, you can enhance your chances of success. Remember that the process is as much about self-discovery as it is about gaining admission to medical school. Stay focused on your passion for medicine, and your dedication will carry you through this exciting chapter of your life.

"Education is the most powerful weapon which
you can use to change the world."
— Nelson Mandela

FIRST EDITION

FROM *Cells* TO SCALPELS

A GUIDE FOR ASPIRING DOCTORS

WRITTEN BY AWARD-WINNING
AUTHOR YASMINE BEN SALMI

CHAPTER 11

CHAPTER 11

Key Takeaways and Inspiring Books to Read

Key Takeaways from "From Cells to Scalpels"

The Journey to Medicine: Becoming a doctor is a unique and rewarding journey, driven by a passion for understanding the human body and improving patient health.

Discover Your Passion: Find your unique passion within the vast field of medicine, exploring various specialities and gaining insight from mentors.

Diverse Specializations: Medicine offers a wide range of specialities, each with its unique challenges and opportunities. Find the one that resonates with you.

The Art of Healing: Remember that medicine is not just science but also an art that involves empathy, compassion, and effective communication with patients.

Trials and Triumphs: The life of a doctor is marked by challenges and rewards.
The journey involves rigorous education, training, and the privilege of caring for patients.

Continuous Learning: Medicine is a lifelong journey of learning and adaptation. Staying current with research and advances is essential.

The Doctor's Duty: A doctor's duty goes beyond diagnosis and treatment. It includes advocating for patient well-being, upholding ethics, and fostering a strong doctor-patient relationship.

Inspiring Books to Read

"Under the Knife: The Inside Story of Surgery" by Dr. Arnold Van de Laar

- **Summary:** Dr. Arnold Van de Laar, a surgeon, provides a fascinating and often humorous look into the world of surgery. He shares real-life anecdotes, historical insights, and the evolution of surgical techniques, offering a unique perspective on a profession that saves lives.

"When Breath Becomes Air" by Dr. Paul Kalanithi

- **Summary:** Dr. Paul Kalanithi, a neurosurgeon, shares his poignant memoir about life, death, and the pursuit of medicine. Diagnosed with terminal lung cancer, Kalanithi reflects on his career, his role as a patient, and the profound meaning of life and mortality.

"The Emperor of All Maladies: A Biography of Cancer" by Dr. Siddhartha Mukherjee

- **Summary:** Dr. Siddhartha Mukherjee, an oncologist, explores the history of cancer, its impact on society, and the relentless pursuit of a cure. This Pulitzer Prize-winning book delves into the scientific, social, and personal aspects of cancer research.

"Complications: A Surgeon's Notes on an Imperfect Science" by Dr. Atul Gawande

- **Summary:** Dr. Atul Gawande, a surgeon and author, offers a thought-provoking collection of essays that shed light on the complexities of medical practice. He explores the uncertainty, challenges, and ethical dilemmas faced by healthcare professionals.

"Being Mortal: Medicine and What Matters in the End" by Dr. Atul Gawande

- **Summary:** In this book, Dr. Atul Gawande delves into the complexities of ageing, end-of-life care, and the role of medicine in ensuring a meaningful and compassionate conclusion to life. He challenges conventional practices and highlights the importance of quality of life in healthcare.

These inspiring books provide valuable insights into the world of medicine, its challenges, and its impact on patients and practitioners. They offer profound stories and thought-provoking perspectives that can enrich your understanding of the field and inspire your journey in medicine.

""The good physician treats the disease; the great physician treats the patient who has the disease."
— William Osler

FIRST EDITION

FROM *Cells* TO SCALPELS

A GUIDE FOR ASPIRING DOCTORS

WRITTEN BY AWARD-WINNING
AUTHOR YASMINE BEN SALMI

NOTES

FROM *Cells* TO SCALPELS

A GUIDE FOR ASPIRING DOCTORS

FROM *Cells* TO SCALPELS

A GUIDE FOR ASPIRING DOCTORS

FROM *Cells* TO SCALPELS

A GUIDE FOR ASPIRING DOCTORS

FROM *Cells*
TO SCALPELS

A GUIDE FOR ASPIRING DOCTORS

_____
_____
_____
_____
_____
_____
_____
_____
_____
_____
_____
_____
_____
_____
_____
_____
_____
_____

FROM *Cells* TO SCALPELS

A GUIDE FOR ASPIRING DOCTORS

_____
_____
_____
_____
_____
_____
_____
_____
_____
_____
_____
_____
_____
_____
_____
_____
_____
_____

FROM *Cells*
TO SCALPELS

A GUIDE FOR ASPIRING DOCTORS

FROM *Cells* TO SCALPELS

A GUIDE FOR ASPIRING DOCTORS

FROM *Cells* TO SCALPELS

A GUIDE FOR ASPIRING DOCTORS

FROM *Cells* TO SCALPELS

A GUIDE FOR ASPIRING DOCTORS

_____
_____
_____
_____
_____
_____
_____
_____
_____
_____
_____
_____
_____
_____
_____
_____
_____
_____

FROM *Cells* TO SCALPELS

A GUIDE FOR ASPIRING DOCTORS

FROM *Cells* TO SCALPELS

A GUIDE FOR ASPIRING DOCTORS

FROM *Cells* TO SCALPELS

A GUIDE FOR ASPIRING DOCTORS

FROM *Cells*
TO SCALPELS

A GUIDE FOR ASPIRING DOCTORS

FROM *Cells* TO SCALPELS

A GUIDE FOR ASPIRING DOCTORS

FROM *Cells* TO SCALPELS

A GUIDE FOR ASPIRING DOCTORS

FROM *Cells*
TO SCALPELS

A GUIDE FOR ASPIRING DOCTORS

FROM *Cells* TO SCALPELS

A GUIDE FOR ASPIRING DOCTORS

FROM *Cells* TO SCALPELS

A GUIDE FOR ASPIRING DOCTORS

FROM *Cells*
TO SCALPELS

A GUIDE FOR ASPIRING DOCTORS

ABOUT
THE AUTHOR

ABOUT THE AUTHOR

Yasmine Ben Salmi, a dynamic entrepreneur, podcast host, and author, has carved an impactful presence in the realms of personal development and cultural connectivity. As the host of the "Life According To Yasmine" podcast, she delves into inspiring narratives that resonate with audiences worldwide.

Yasmine is the visionary founder of The Choice Is Yours Publishing House and the innovative dog walking service, "Woof-Woof your dog is here." Her commitment to family and community is exemplified by her recognition during Chelsea FC's Edge of The Box 6th Anniversary celebration and her facilitation of the signature family workshop, "Dreaming Big Together - Mamas Secret Recipe," at The Hub Chelsea FC & Virgin Money.

In the sphere of educational initiatives, Yasmine's family has participated in masterclasses at Brunel University London, where her youngest brother, Amire, achieved the remarkable distinction of being the youngest-ever honorary STEM Ambassador in the university's history. An accomplished author and entrepreneur, Yasmine has founded the Mother and Daughter Connect Collection and Lovepreneur. She hosts the motivational program "The Choice Is Yours - Your Thinking C.A.P For Living & Loving Life" at Virgin Money Lounge, inspiring positive transformations.

Yasmine's brand collaborations include campaigns with Sainsbury's, Legoland, Warner Bros, Sony, and Made for Mums. Her literary contributions include books such as "The Choice Is Yours: Your Thinking C.A.P for Living & Loving Life Part 2" and "Can I Ask You A Question Doctor?: Neurology Edition with Mr Chidiebere Ibe." with an extensive portfolio in motivational programs, brand collaborations, and a compelling array of literary works, Yasmine Ben Salmi continues to make a significant impact on diverse audiences. Her multifaceted contributions exemplify a dedication to empowering others and creating positive change in both the personal and professional spheres.

Yasmine is proud to be embarking on her A Levels in Phycology, Chemistry and Biology at Highgrove Education with the incredible Ms Heather Rhodes as her principal.

FROM *Cells* TO SCALPELS

FIRST EDITION

WRITTEN BY AWARD-WINNING
AUTHOR YASMINE BEN SALMI

LET'S STAY CONNECTED

LET'S STAY CONNECTED

in   Yasmine Ben Salmi

◉   @AuthorYasmineBenSalmi

✉   Info@DreamingBigTogether.com

" IN THE DANCE OF LIFE, MEDICINE IS THE RHYTHM
OF HEALING. AS YOU STEP INTO THIS SYMPHONY,
LET COMPASSION GUIDE YOUR STEPS, RESILIENCE
BE YOUR MELODY, AND EACH PATIENT ENCOUNTER
BECOME A NOTE IN THE BEAUTIFUL COMPOSITION
OF A MEANINGFUL EXISTENCE.

"As you embark on this extraordinary journey in medicine, may your compassion be your guide, your knowledge be your armor, and your heart be your greatest tool. Remember, in each heartbeat, in each diagnosis, and in each triumph, you carry the potential to make a lasting impact.

Your journey is not just a path; it is a testament to the healing power of dedication and humanity. May your story be written with the ink of hope and the pages of countless lives touched. The world awaits the imprint of your scalpel and the warmth of your healing touch. Your journey begins, and within it lies the promise of a profound and purposeful legacy."

BONUS CONTENT

ASGBI
Association of Surgeons of
Great Britain and Ireland

I am proud to be a member of the Association of Surgeons of Great Britain and Ireland.

**British Association
of Black Surgeons**

I am proud to be a member of the British Association of Black Surgeons.

My first step into the medical world.

As young as 7 years old, I had the dream of becoming a plastic surgeon, but I never thought of how quickly it would become a reality. At the age of 14 years old, I co-authored a book with Mr Chidiebere Ibe who is famously known as the "Black Foetus Man", he created the first ever medical illustration for a black foetus. He had a DREAM just as I have to change the world in a positive way, he illustrated the black fetus and other medical illusions so that doctors could diagnose patients according to their skin colour. It is so fascinating how he DREAMED of impacting people in a way no one would have ever expected.

I am proud to say that I have published a scientific article with her mentor Detina Zalli founder of We Speak Science, who's a lead professor from Harvard, Oxford and Cambridge: https://oxfordacademy.io/overview-of-reconstructive-plastic-surgery-advantages-and-disadvantages/ you can read my article on the following pages.

Overview Of Reconstructive Plastic Surgery; Advantages And Disadvantages

Introduction: In this article, we will describe the advantages and disadvantages of reconstructive plastic surgery. The first part of the article will introduce reconstructive plastic surgery, and why it is done. How reconstructive plastic surgery has advanced over time, and how it has impacted people. In the second section of the article, we'll talk about the disadvantages of reconstructive plastic surgery as well as what to hope for in this field in the future.

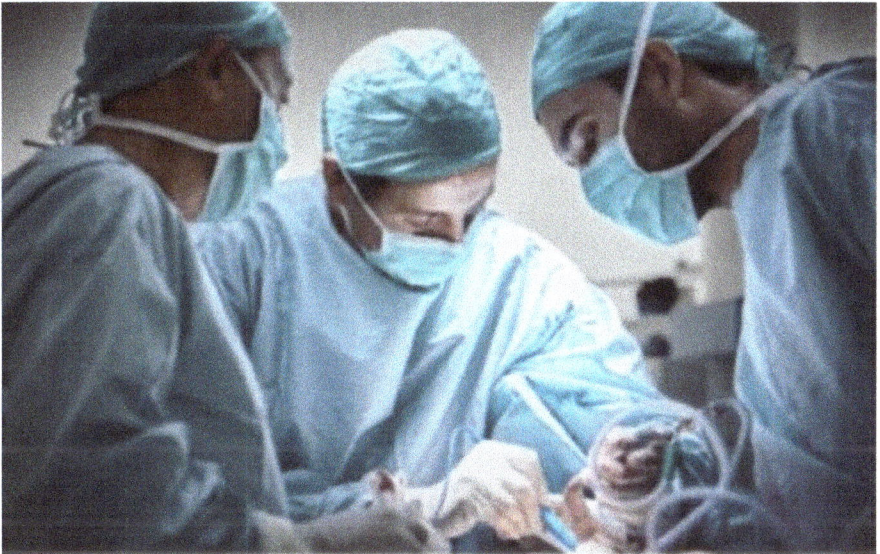

Image Source / Getty Images

Introduction to Reconstructive Surgery

There has been a long history of human beings aiming to succeed in self-improvement since the beginning of time. This may explain why plastic surgery could possibly be one of the world's oldest healing practices. It has been documented that over 4,000 years ago, facial injuries were corrected with surgical means. As early as 800 B.C. Physicians in Ancient India used skin grafts to reconstruct the body. Overall progress in plastic surgery, like most medicine, was slow over the next few thousand years, as techniques used in India were introduced to the West and then subsequently refined and adapted for new applications. However, there was progress made in medicine during the Greco-Roman period, and that progress was documented in ancient texts which were disseminated over time throughout civilization. However, Reconstructive Surgery was first introduced in medicine during the First World War in 1917 by Surgeon Harold Gillies, as there was an increase in drastic facial injuries due to the War. This urged Surgeon Harold Gillies to come up with a new method of facial reconstructive surgery. Because of his great discovery, his works marked the dawn of plastic surgery as we know it today.

What is Reconstructive Plastic Surgery, and why is it done?

In my opinion reconstructive plastic surgery is very impactful in the way it has changed people's lives. Here is a brief explanation of what Reconstructive Plastic

Surgery is, and why it is done. From the top of the head to the tip of the toe, and from newborn babies to the very elderly, reconstructive surgery is used to treat a wide range of conditions. Reconstructive surgery is all about repairing people and restoring function. It is performed to repair and reshape bodily structures affected by birth defects, developmental abnormalities, trauma/injuries, infections, tumours and disease. Using a wide range of reconstructive techniques, plastic surgeons mend holes and repair damage primarily through the transfer of tissue from one part of the body to another. Their main aim is to restore the body, or the function of a specific part of the body, to normal. However, plastic surgeons carrying out reconstructive surgery also try to improve and restore appearance. Wherever possible they attempt to minimise the visual impact of the initial wound or defect, and the impact of the surgery itself. Cosmetic surgery is an extension of reconstructive plastic surgery in which the main functional gain to be expected is an improvement in appearance.

A brief explanation of how Reconstructive Plastic Surgery advanced over time

Image Source / Getty Images

Over the last few years reconstructive plastic surgery has been a game changer in the medical world. Especially when exploring the fact that reconstructive surgery first started off as skin grafts in ancient India to reattaching limbs. Here is a quick overview of how reconstructive plastic surgery has advanced over time. Reconstructive plastic surgery was used to enhance what was already there, or to solve common defects within people, but when people started to understand what you could really do with reconstructive plastic surgery, that was when cosmetic surgery was born. People started to understand that you could not only enhance your looks, but you can also change them. Plastic surgery has become synonymous with the quest for youth and beauty, albeit with varying degrees of success. But the field has, for centuries, been driven by medical necessity — and it has nothing to do with plastic. The discipline derives its name from the Greek word "plastikos" — to mold or give form. And while the idea of perfecting yourself surgically is a relatively recent phenomenon, there is evidence of reconstructive surgery going back to antiquity. The oldest known procedures appear in an ancient Egyptian medical text called the "Edwin Smith Papyrus." Thought to be an early trauma surgery textbook (and named after the American Egyptologist who purchased it in 1862), the treatise contains detailed case studies for a variety of injuries and diagnoses. As well as showing how the Egyptians treated wounds and bone fractures, the papyrus revealed a suggested fix for nasal injuries: manipulating the nose into the desired position before using wooden splints, lint, swabs and linen plugs to hold it in place. The Egyptians occasionally used prosthetics, too: In 2000, an ancient mummy was found to have a prosthetic toe that may have aided the woman's walking, according to researchers who tested replicas of the toe on modern-day volunteers. Whether these procedures can be considered types of plastic surgery is a matter of historical debate, according to Justin Yousef, whose research on the topic was recently published in the European Journal of Plastic Surgery. It is in India, in fact, that historians have found "the first description of reconstruction proper," he said in a phone interview. By the 6th century B.C., physicians in India were carrying out procedures not dissimilar to a modern-day cosmetic rhinoplasty. In a detailed compendium called "Sushruta Samhita," the Indian physician Sushruta — who is sometimes called the father of plastic surgery — outlined a remarkably advanced technique for skin grafts. As in Egypt, the procedure involved repairing noses. But according to Yousef, patients' motives were, in a sense, cosmetic.

How has Reconstructive Plastic Surgery helped people?

Reconstructive surgery is a technique that heals your body after an illness or injury or repairs birth defects. It not only heals your body, but also gives you comfort so that you feel comfortable in your own body. Some of my Fav reconstructive procedures to watch in my spare time include, face reconstruction, hand reconstruction using skin graphing techniques and breast reconstruction using harvesting sites on the body. When watching videos, where people was either born with a defect that didn't allow them to live comfortably, or whether they developed illnesses or was in an accident, I find it fascinating to see how much reconstructive surgery is able to change people's life for the better.

Here is a life-saving Reconstructive Plastic Surgery that was performed on a 20-year-old from Tanzania, performed by Neuroplastic Surgeon Chad Gordan and neurosurgeon Judy Huang at John Hopkins Hospital.

Life-saving surgery in Tanzania after a two-story fall left Dennis, then age 20, alive, but with 45% of his skull missing. Stalled in his recovery due to incomplete wound healing and underlying bone infection, Dennis turned to Johns Hopkins neuroplastic surgeon Chad Gordon and neurosurgeon Judy Huang, who devised a two-surgery approach. The first procedure addressed the infected bone segment that was causing severe deformity, and the second offered a transformative solution: a custom implant to shape Dennis' head, replace the missing skull and protect his brain. Now fully healed, Dennis can play soccer and return to school.

What are some of the disadvantages of having Reconstructive Plastic Surgery?

When thinking about having Reconstructive Plastic Surgery, it is very important to keep in mind that there are risks involved, just like any other procedure, operation or surgery. Some of the Risks include: **Infection, Risk of Death, Excessive Bleeding, Bruising, Difficulty in wound healing, anaesthesia problems, surgery problems, patients being dissatisfied with their results.**
Studies have shown that those who have undergone Reconstructive Plastic Surgery some of the complications may increase if a patient.
Smokes, has connective tissue damage, has decreased circulation at the site of the surgery, has HIV (Human immunodeficiency virus), has impaired immune system, has poor nutritional habits.
If you do worry that you are feeling discomfort with the surgery site, please do consult your doctor.

What can we expect from Reconstructive Plastic

Surgery in the future?

Some things in the future of plastic surgery are relatively easy to predict such as an increase in popularity of aesthetic surgery and, which procedures will set the pace. When we look at new technologies and advancements it's more difficult to forecast what will become mainstream in the future. What is certain though, is that plastic surgery will continue to develop and be shaped as new advancements and innovations are introduced and accepted. Cosmetic surgery trends indicate that the demand for plastic surgery will continue to grow as it becomes more advanced, less invasive and more affordable. New technologies and innovative techniques will be pioneered to improve the quality of procedures even further. In today's world, many people want to look and feel their best. Plastic surgery can help achieve this and delay the ageing process. In the future this desire to look young and attractive may become even more commonplace as people work and live longer. There are some plastic surgeries that can be predicted to continue in their popularity. For example, the number of liposuction procedures being performed is growing in the UK and US. As obesity and overweight levels are increasing in these populations it's reasonable to predict that liposuction will remain a top plastic surgery procedure for the foreseeable future.

Now and in the future, it's thought that the trend is for natural and incremental plastic surgery changes. Non-surgical treatments have also become more popular over recent years and it's expected for this to carry on. Patients will continue to combine these procedures and seek completely non-invasive options to manage their facial ageing and to provide them with the body features they desire.

Stem cell technology is an exciting concept for plastic surgeons. Stem cells can be harvested to regenerate cells and tissues in the human body. In the future stem cell technologies may be combined with tissue engineering to grow new body parts as required. These structures such as skin and ears will be grown in the laboratory and implanted to restore form and function. Stem cells could also potentially be re-injected into areas as a filler that is more compatible and longer lasting than current fillers.

Areas where research is currently being undertaken include:

- wound care – use of skin substitutes (composed of

living cells grown in a laboratory) to heal wounds.

- scar treatment to improve the healing of scars.
- hand and face transplantation. Burn care – generation of the dermal (innermost skin) layer following burn injury.
- breast reconstruction – regeneration of new tissue layers over implants in breast cancer survivors.
- nerve regeneration – regeneration of nerves and restoration of optimal function after nerve injury
- wound care – use of skin substitutes (composed of living cells grown in a laboratory) to heal wounds.
- scar treatment to improve the healing of scars.
- hand and face transplantation.

A patient's extra fat tissue can be transferred and regenerated in other parts of their body. This revolutionary technique has been used in plastic surgery on the breasts, buttocks, face and hands. However, this is currently only an option for patients who have sufficient fat on other areas of the body.

Stem cells can become any type of tissue such as fat and it may become possible to grow significant amounts of the patient's own fat in the laboratory that can then be transferred to their breasts, buttocks, or other areas. One day this could make implants a thing of the past.

Treatments may be developed to modify programmed cell death, the breakdown of collagen and other skin and soft-tissue changes that cause people to develop an aged appearance.

There are already some radical new techniques for fat reduction that don't require incisions. Current procedures and those in the pipeline include fat-freezing and fat-burning that will destroy fat cells and allow problem areas to be specifically targeted.

In the last decade, the number of procedure types has increased. It's expected that plastic surgery options will continue to expand allowing for more bespoke treatments.

There is an increased in how individuals react differently to certain treatments or products. In the future, individual cosmetic treatments may be tailored by specific patient profiles to achieve the best possible results and it may be possible to determine if a patient will scar more than another patient.

The future of plastic surgery is certainly a fascinating and exciting area.

Conclusion:

In conclusion, reconstructive surgery is the new method for enabling people to live freely in their own skin, however, it comes with an additional element of risk as all surgeries include some degree of risk. Reconstructive plastic surgery, on the other hand, enables people who have had birth deformities in the past to live their lives without worrying about how they look. Having said that, I expect to see a lot more people striving to change medicine in the future, no matter what obstacles they face. There is always a method as long as we can find the solution, whether it be now or in the near future.

———————

References:

1. The history of plastic surgery over time https://www.verywellhealth.com/the-history-of-plastic-surgery-2710193
2. The founding father of facial reconstructive plastic surgery Surgeon Harold Gillies https://www.nam.ac.uk/explore/birth-plastic-surgery
3. What is Reconstructive Surgery, and why is it

done? https://www.bapras.org.uk/public/patient-information/reconstructive-surgery

4. How has Reconstructive Surgery evolved over time? https://edition.cnn.com/style/article/plastic-cosmetic-surgery-history-scn/index.html

5. Here is a life-saving Reconstructive Plastic Surgery that was performed on a 20-year-old from Tanzania, performed by Neuroplastic Surgeon Chad Gordan and neurosurgeon Judy Huang at John Hopkins Hospital. https://www.hopkinsmedicine.org/plastic_reco appts/patient_stories/

6. Disadvantages of having Reconstructive Plastic Surgery https://stanfordhealthcare.org/medical-treatments/r/reconstructive-plastic-surgery/complications.html

7. What can we expect from Reconstructive Surgery in the future? https://www.northdownshospital.co.uk/news/the-future-of-cosmetic-surgery

Share: Ⓕ Ⓨ

Thank you for taking the time to read my scientific article.
Please visit:

https://oxfordacademy.io/overview-of-reconstructive-plastic-surgery-advantages-and-disadvantages/
and click on the article tab to read my entire article.

FROM *Cells* TO SCALPELS

FIRST EDITION